Parents:

Learning to tell time is an exciting achievement for children. They can tell when it's time for mother and/or father to get home from work, when their favorite television progams are on, when it's time to leave for a special outing, etc. It is a skill we take for granted since we learned how to do it so long ago. Some children find it easier to learn than others. Be patient and provide opportunities for your child to find out what time it is. The activities in this book will help you and your child reach this goal.

Go on a "Clock Search." (page 2)

Walk around the house with your child. Have him/her point out all the places that tell time. Point out places your child overlooks such as the clock on the microwave oven or the VCR. Have your child draw the time pieces he/she finds on page 2 of this book.

Make a paper plate clock (page 3)

You will need to provide your child with these items to use in making a clock face:

- 2 paper plates (or a cardboard circle and some scraps)
- a large paper fastener
- scissors
- glue
- sharp pencil

Steps to follow:

1. If your child can use scissors, have him/her cut out the clock face and the hands on page 3. If not, cut the pieces out yourself.
2. Help your child glue the clock face to one paper plate and the hands to the second paper plate. After the glue has dried, cut the hands out again (the paper plate will give them added strength).
3. Use the pencil to poke a hole in the clock and hands where marked with a small circle. You will need to do this for your child.
4. Place the hands on the clock face and attach them with the paper fastener. Move the hands around a few times to be sure they are moving freely.

Your child will need this clock several times while doing the activities in this book.

1   Understanding clocks

Parents: Read the directions for a clock search on page 1.

I'm Going on a Clock Search

Look.
Draw.

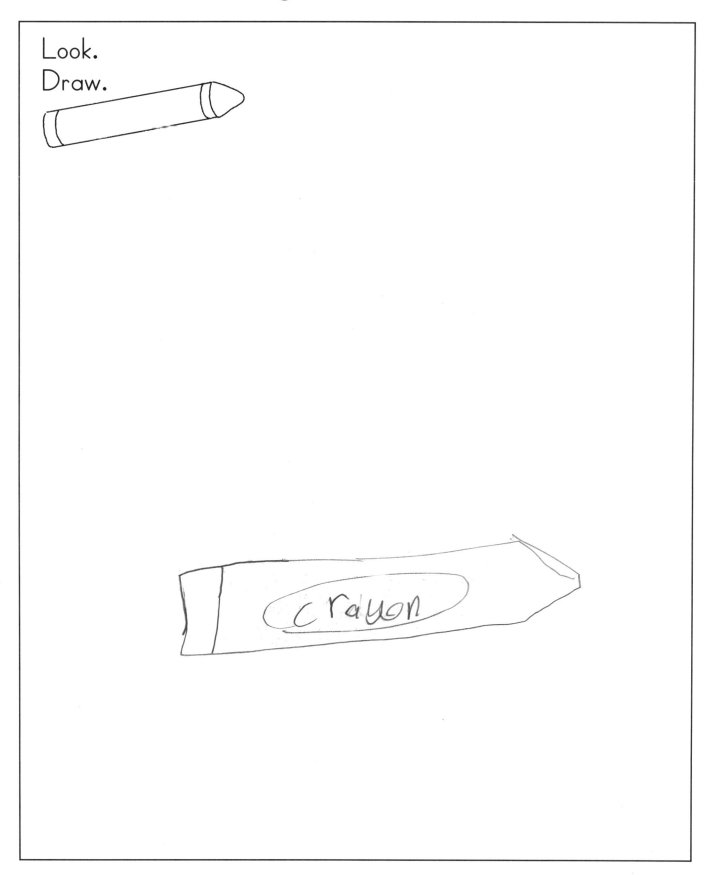

2 Understanding clocks

Parents: Read the directions on page 1 for making this clock.

My Own Clock

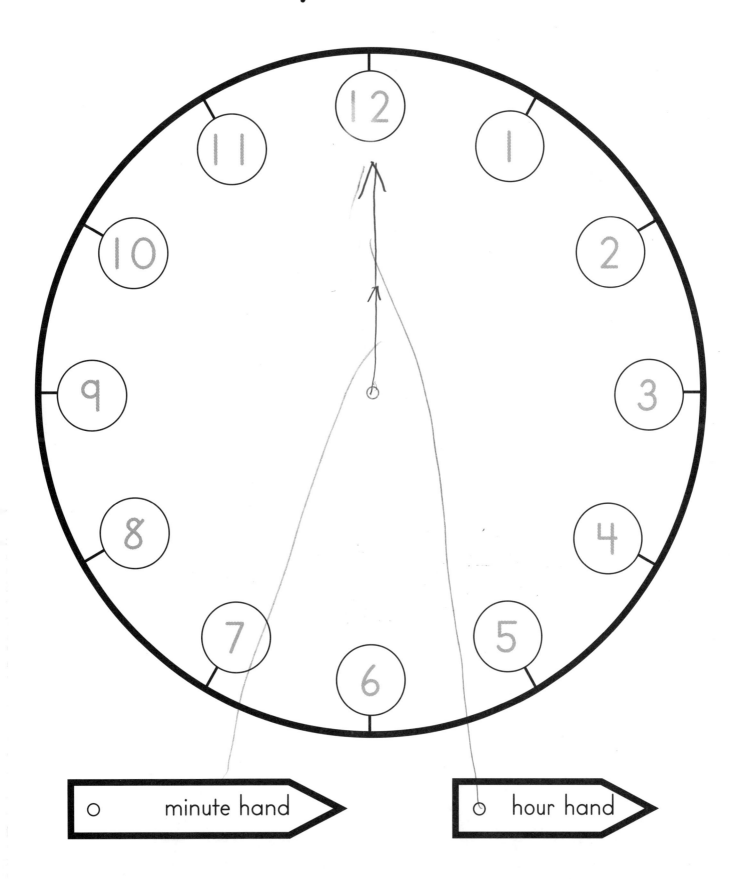

3 Understanding clocks

Parents: Explain to your child that when the big hand is on the 12, we look at the little hand to see what hour it is. Practice with the paper plate clock, having your child put the long hand on the 12 and the little hand pointing to each number in turn. Have him/her say "It is _3 o'clock_." and so on as the little hand is moved. Then look at the clocks on this page. Have your child tell you the hour.

What time is it?

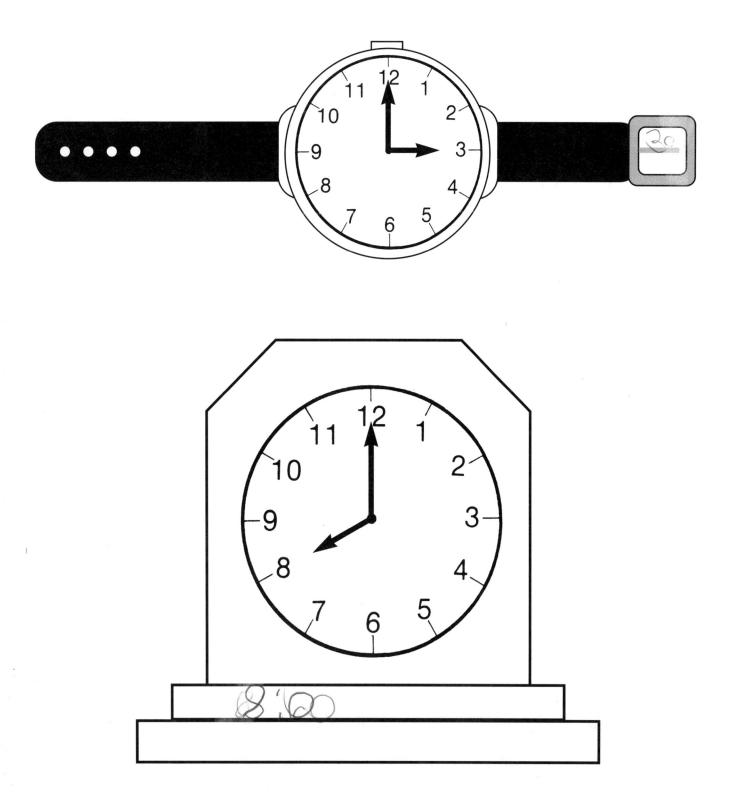

5   Telling the hour

What is the hour?

Circle.

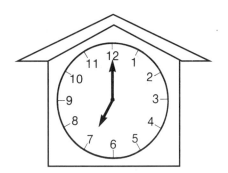

2 o'clock
4 o'clock
(7 o'clock)

(4 o'clock)
12 o'clock
5 o'clock

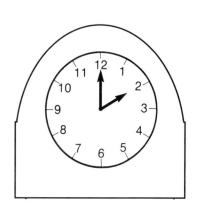

6 o'clock
(2 o'clock)
12 o'clock

8:00
7:00
(9:00)

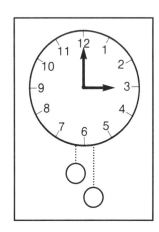

(3:00)
6:00
4:00

1:00
12:00
(11:00)

Match

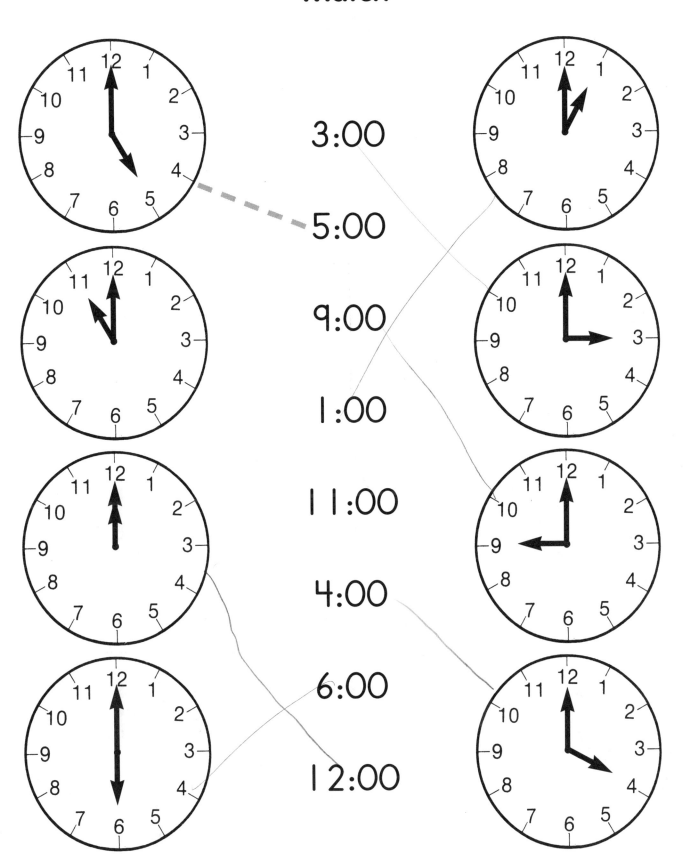

7 Telling the hour

Write the time.

8 Telling the hour

At the Zoo

1. Bob went to the zoo at 2:00.
   He went home at 3:00.
   He was at the zoo ____ hours.

2. Jill went to the zoo at 3:00.
   She can stay 3 hours.
   She must go home at ____ o'clock.

3. Carlos was at the zoo 2 hours.
   He came at 10:00.
   He went home at ____ o'clock.

4. Ann left home at 8:00.
   It took 1 hour to get to the zoo.
   She got to the zoo at ____ o'clock.

Half Past the Hour

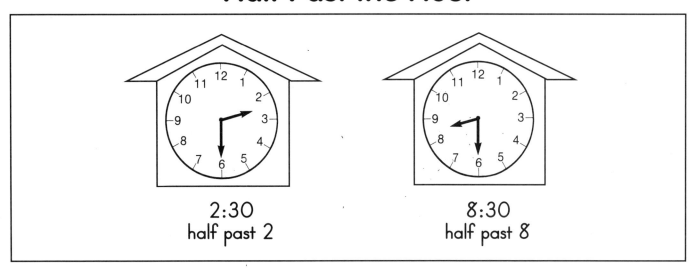

Match.

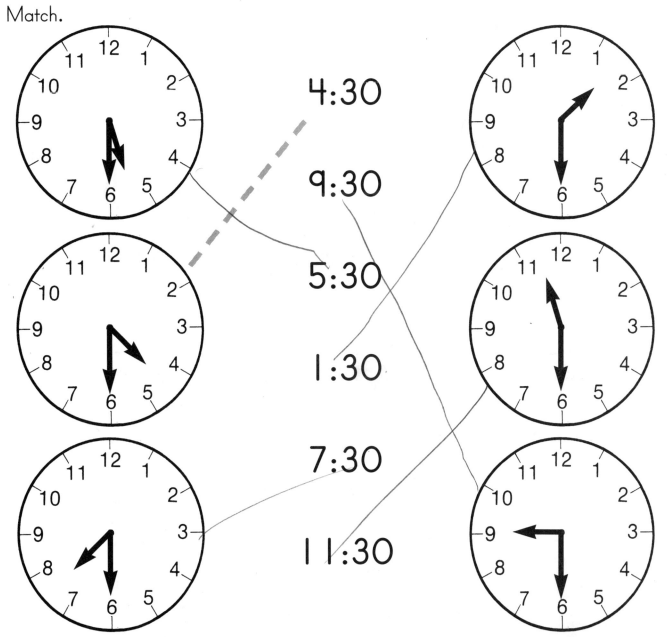

11 Telling the half hour; understanding "half past"

Write the time.

12 Telling the half hour

15 Minutes After the Hour

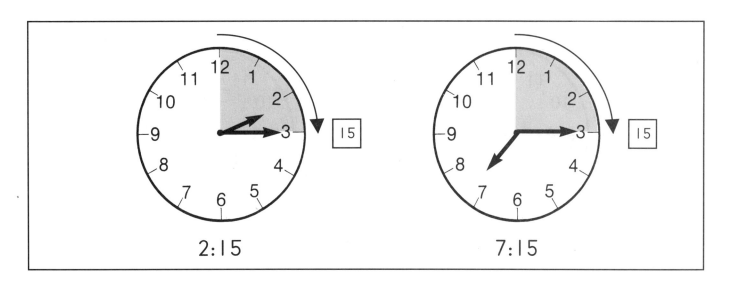

15 Telling 15 minutes past the hour

45 Minutes After the Hour

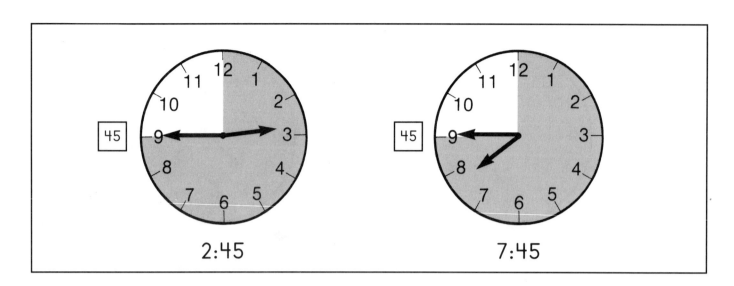

16 Telling 45 minutes past the hour

Match.

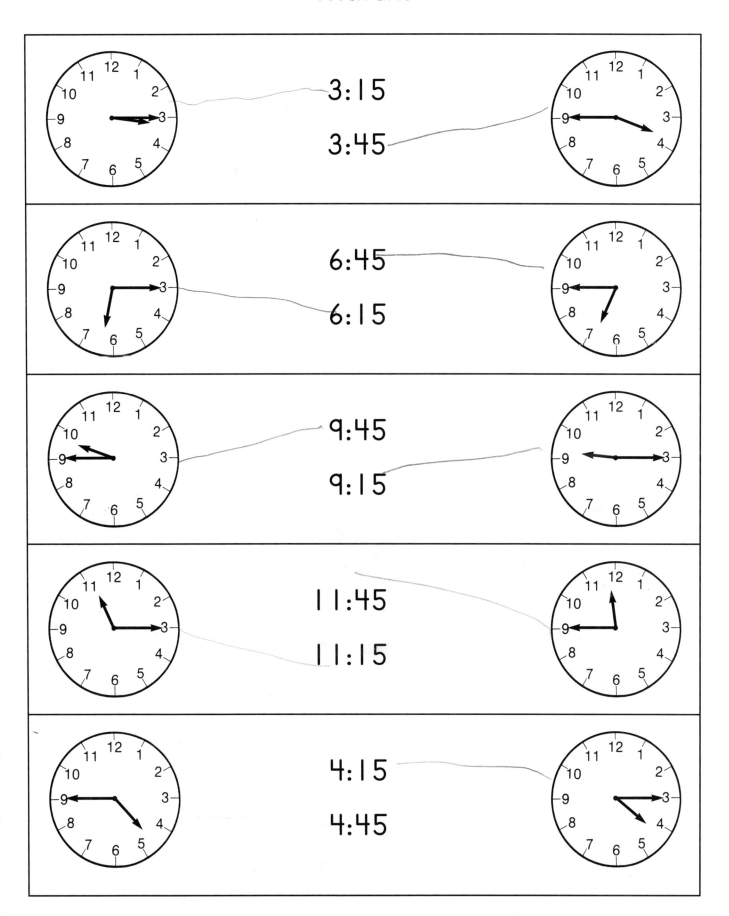

17 Telling 15 and 45 minutes past the hour

Match.

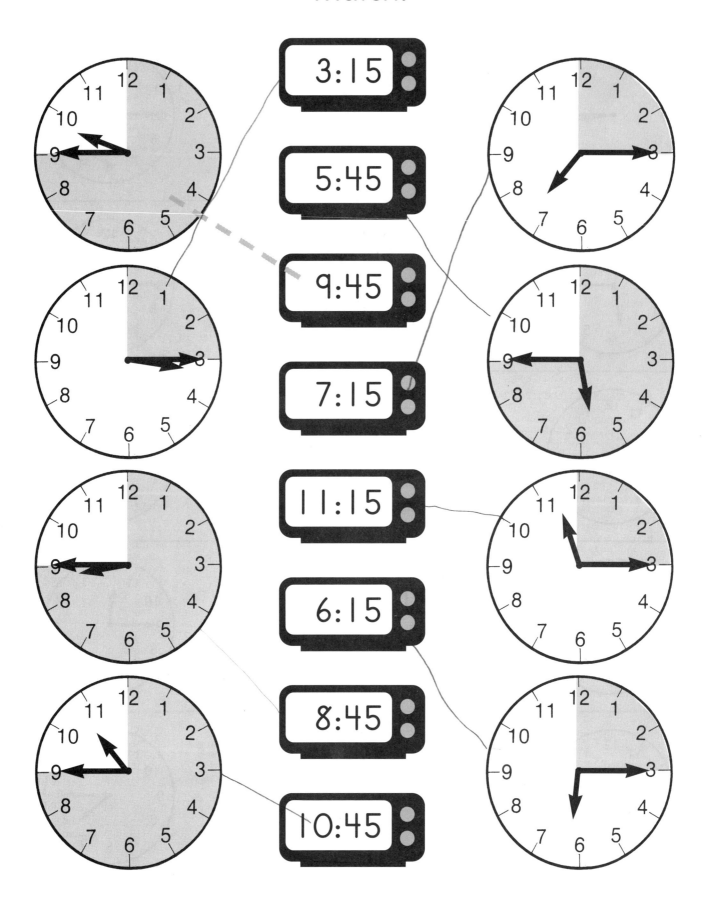

18 Telling 15 and 45 minutes past the hour

Trace the numbers.

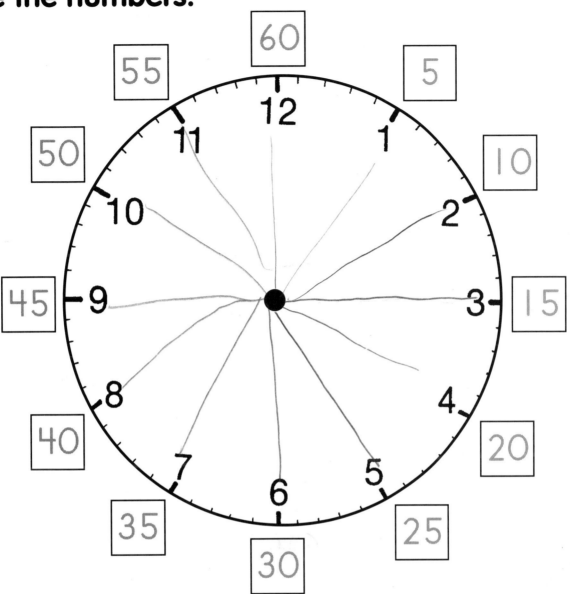

Write.

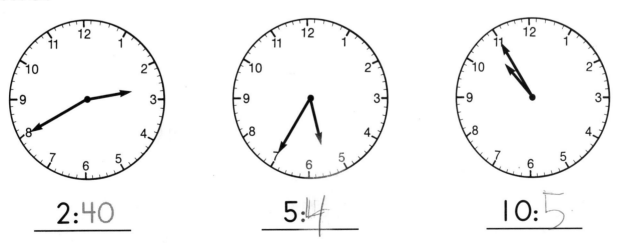

2:40   5:4   10:5

Circle the correct time.

Count the minutes.

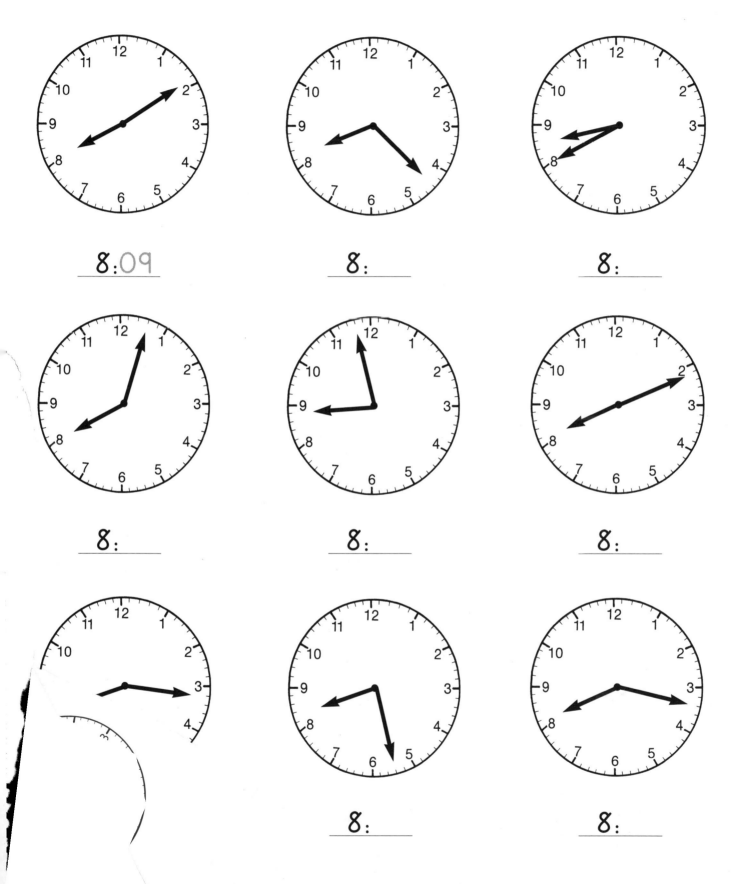

25 Telling time to the minute

Match.

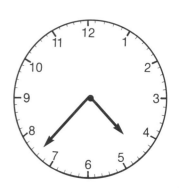

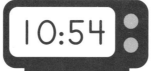
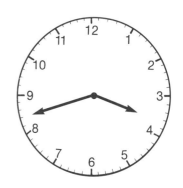

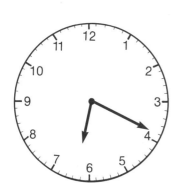

Telling time to the minute

Put the minute hand on the clock face.

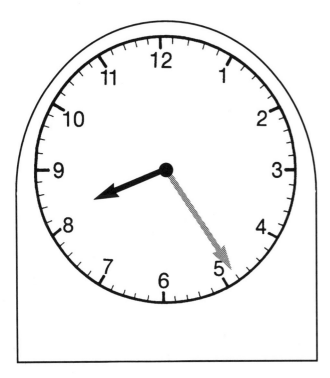

8:24

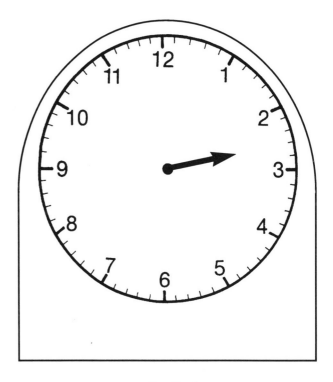

2:26

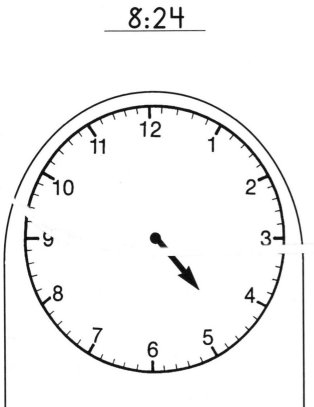

4:38

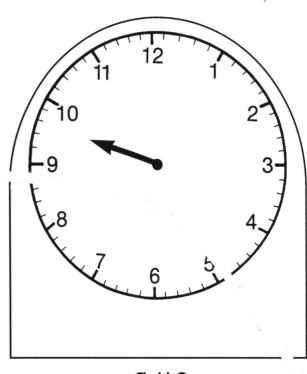

9:40

27 Telling time to the minute

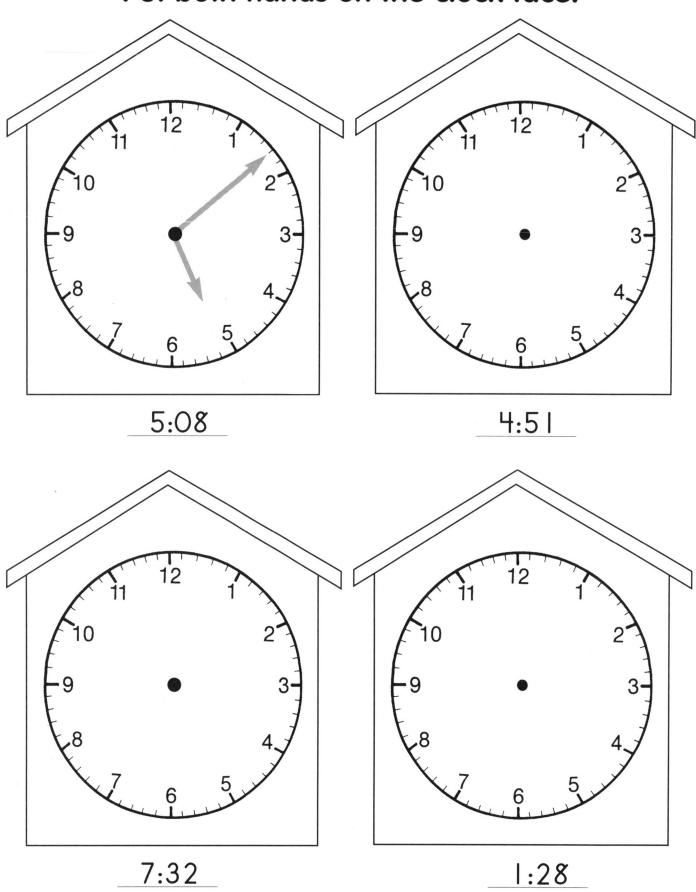

Pennies

Color the 🪙 brown.

How many 🪙? _____

31 Recognizing pennies

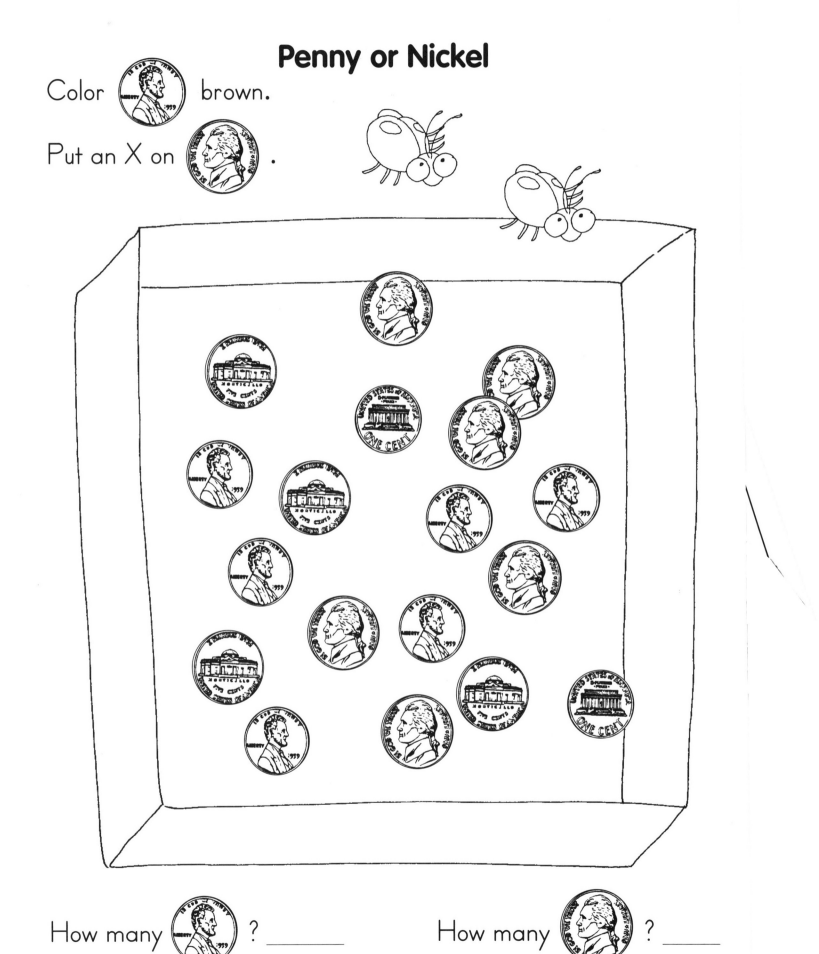

Count the Coins

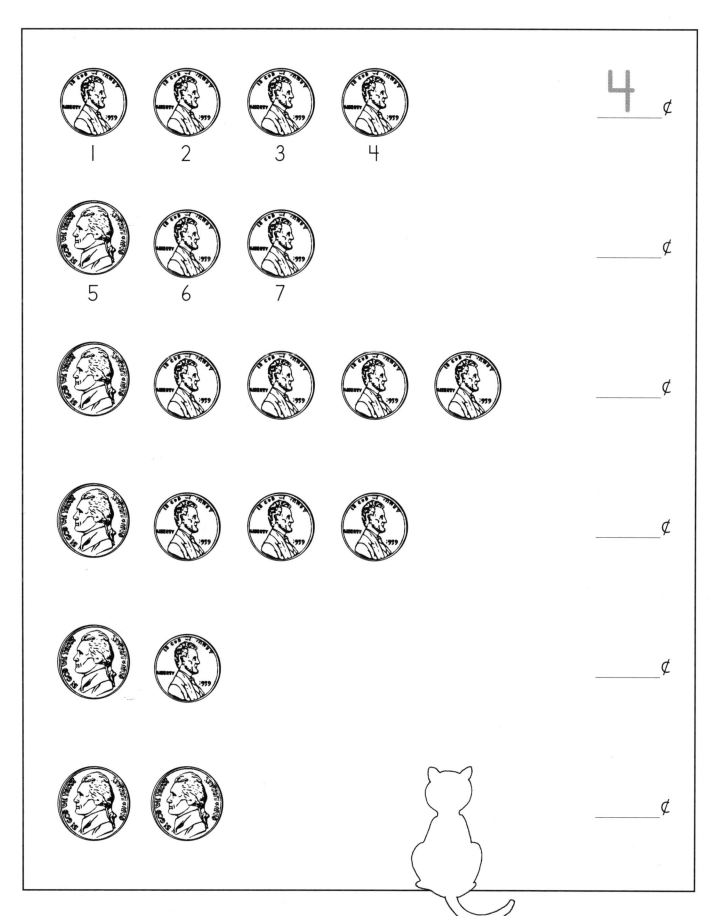

33 Adding change

The Piggy Bank

How much is in the ?

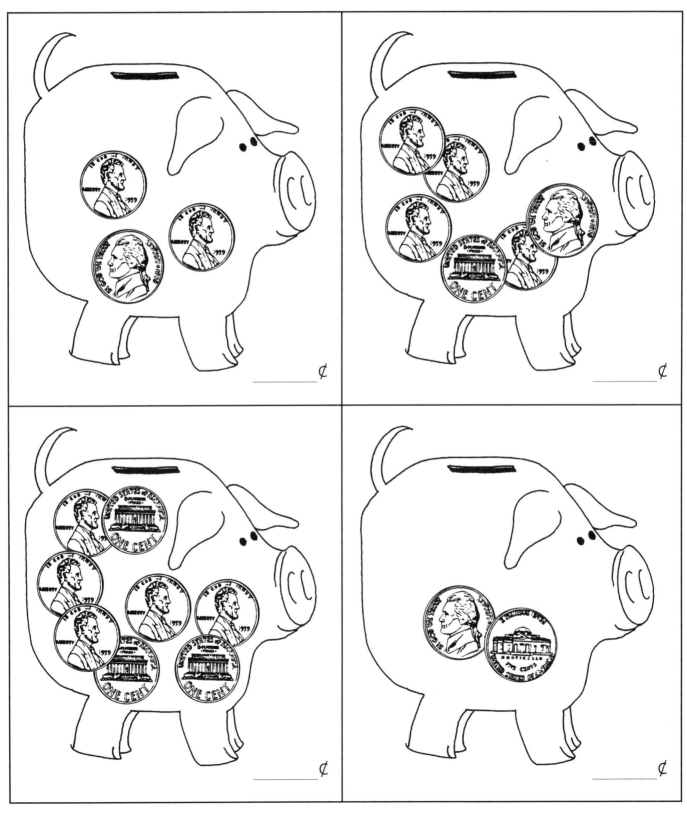

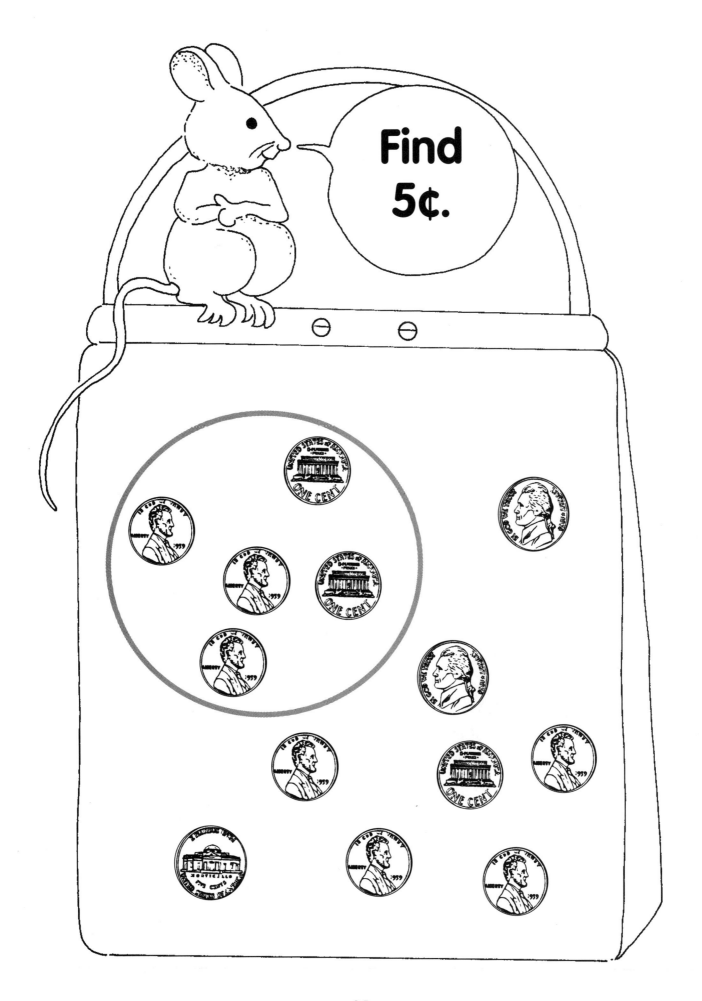

35 Understanding money word problems

For Sale

 3¢
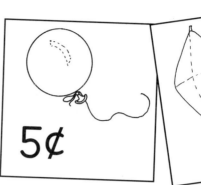 5¢
 8¢

I have:	Can I buy?	yes no
1 nickel, 3 pennies	balloon	Yes
6 pennies	kite	_____
1 nickel, 6 pennies	pencil	_____
2 nickels	balloon and pencil	_____

36 Understanding money word problems

Will I get change back?

It costs:	I have:	Will I get change?
15¢	(4 nickels, 3 pennies)	(yes) no
21¢	(4 nickels, 6 pennies)	yes no
12¢	(2 nickels, 4 pennies)	yes no

37 Making change; understanding money word problems

Fives

Count by 5s.

by 5s dot-to-dot

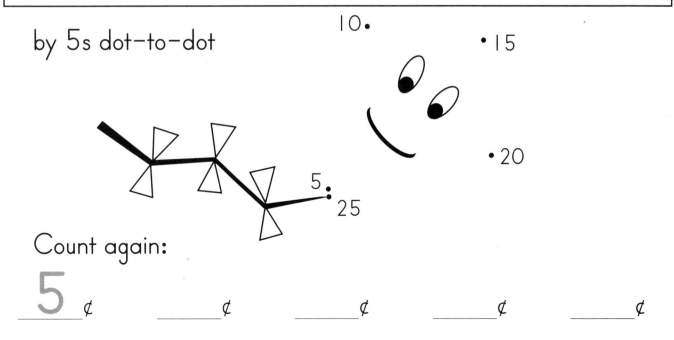

Count again:

5 ¢   ____ ¢   ____ ¢   ____ ¢   ____ ¢

Find the Dimes

Circle the 🪙.

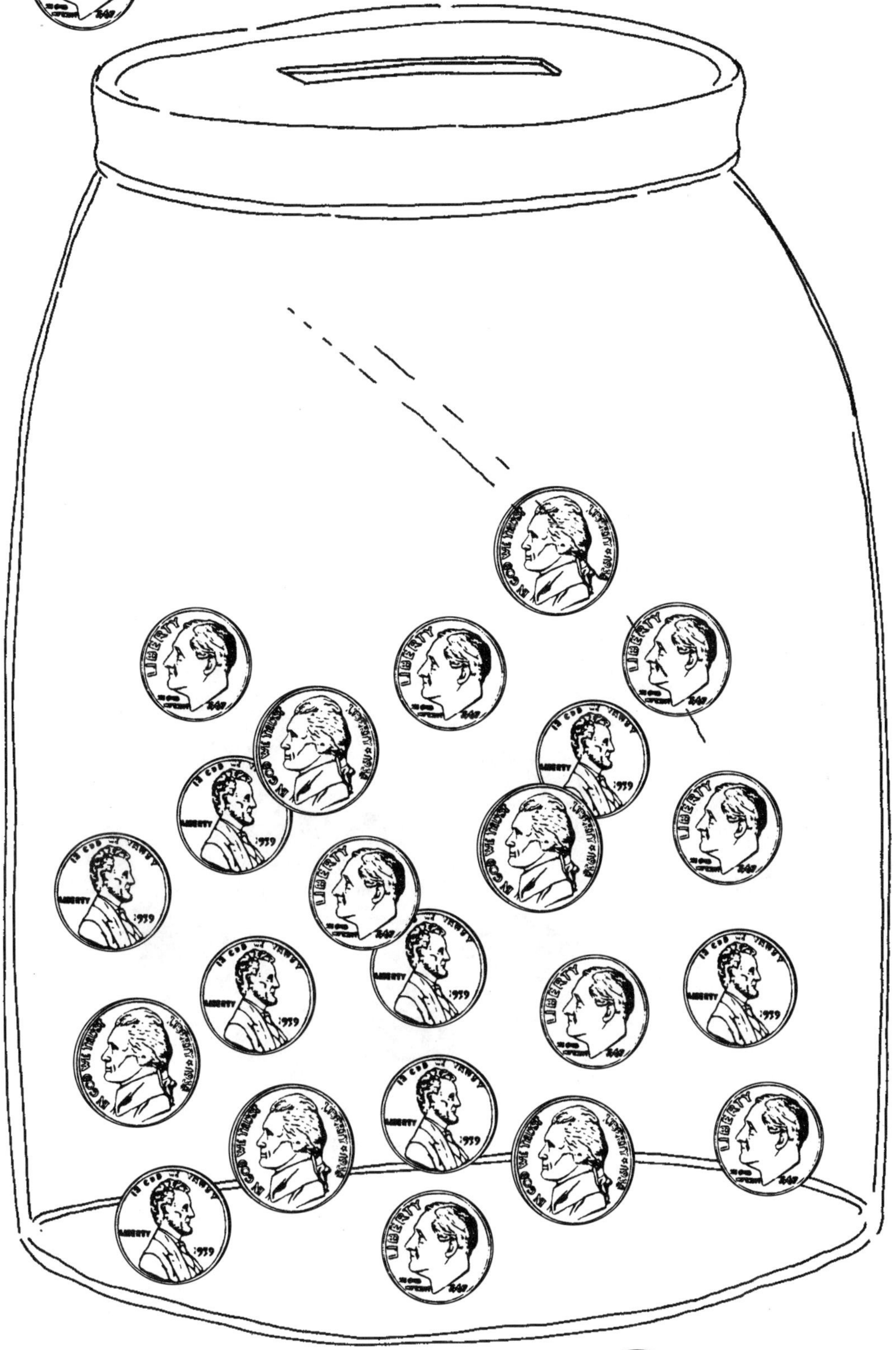

How many 🪙? _____

39 Recognizing dimes

How many ways can you make 10¢?

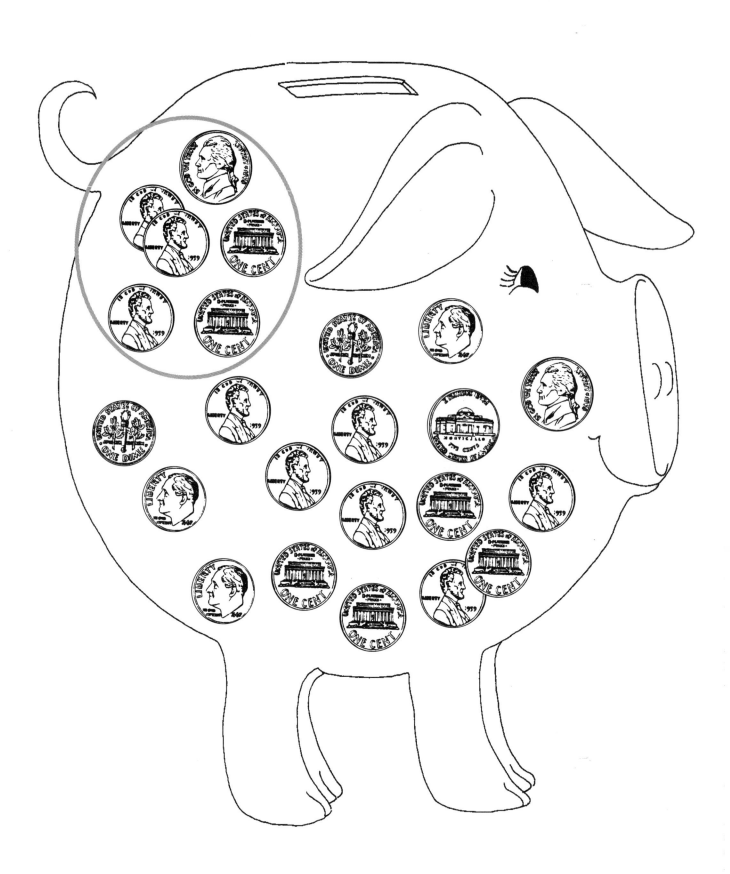

Circle what you need.

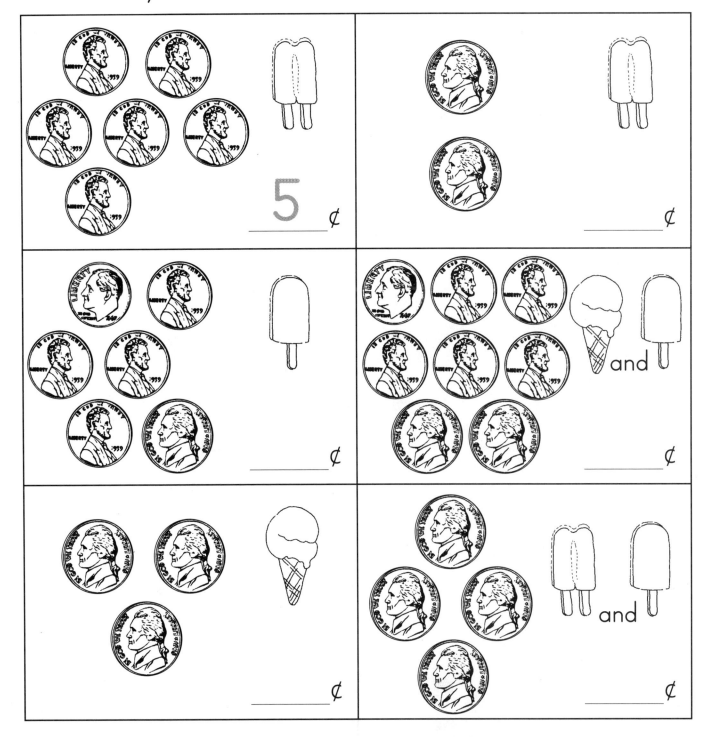

41 Understanding money word problems

Tens

Count by 10s.

_____¢   _____¢   _____¢   _____¢   _____¢

30¢ •

20¢ •   .40¢

Fido

10¢ •   .50¢

Count again:

_____¢   _____¢   _____¢   _____¢   _____¢

42 Adding dimes

Find the Sum

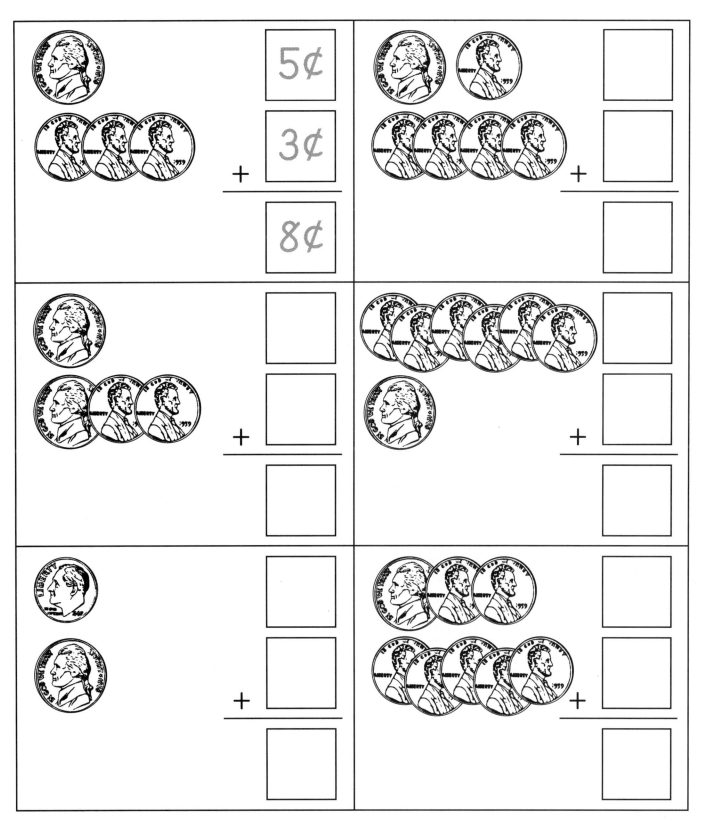

Count the Money

dime   nickel   penny
10¢   5¢   1¢

___ ¢	___ ¢	___ ¢
 ___ ¢	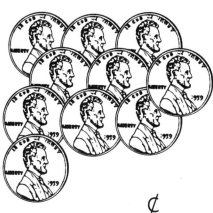 ___ ¢	 ___ ¢
 ___ ¢	 ___ ¢	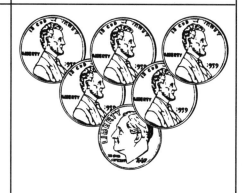 ___ ¢

44   Recognizing coins; adding change

Count On

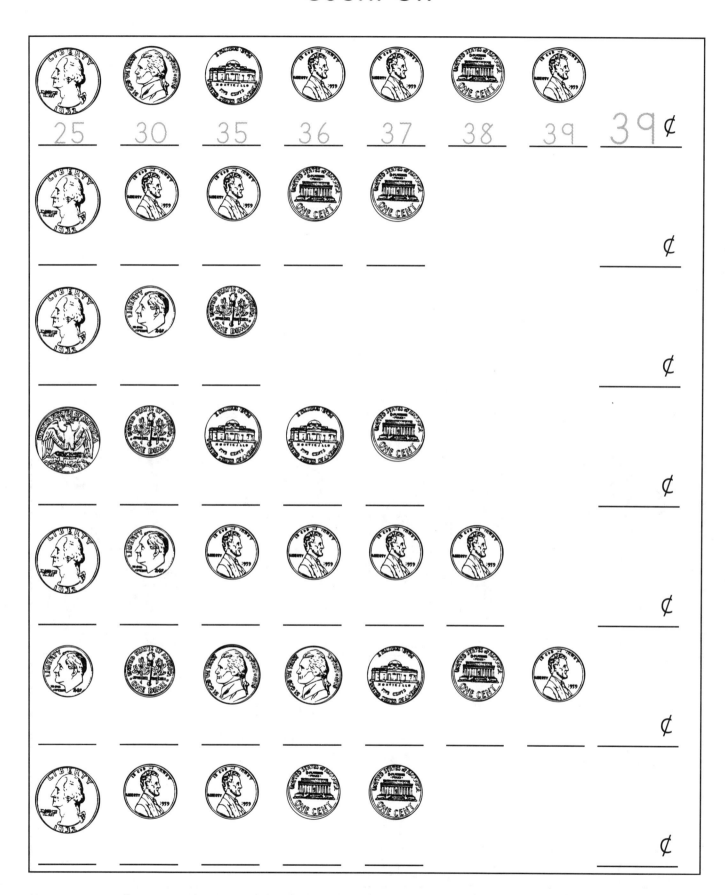

45 Adding change

I Love Cookies!

Sale
10¢   5¢

Circle how much money I need to buy these cookies.

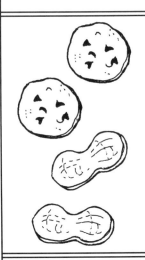		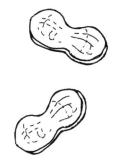	
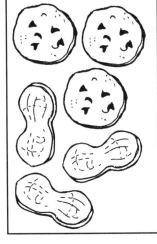		I spent 35¢. How many cookies did I buy? _____ _____	

 25¢
 10¢
 5¢
 1¢

Circle the coins to make these amounts.

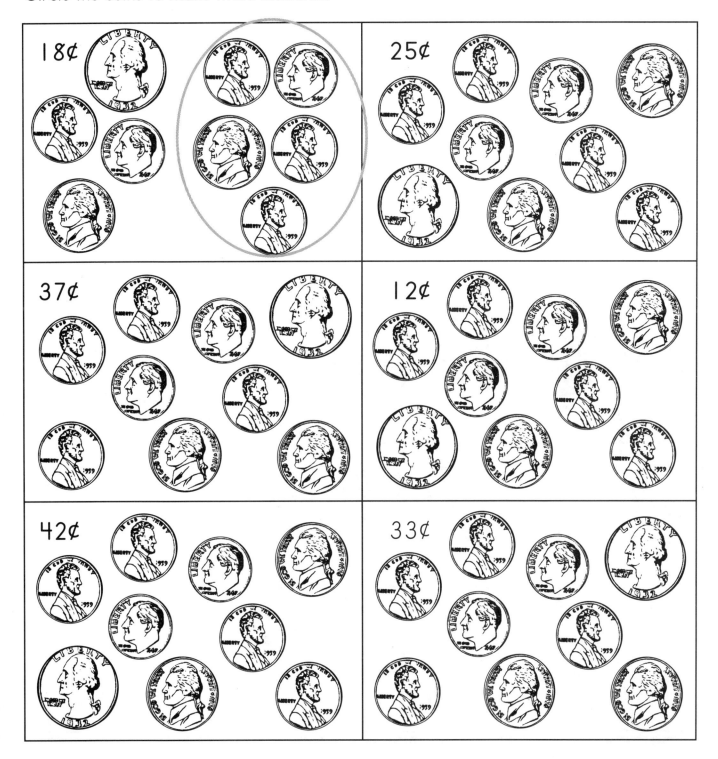

47 Matching coins with amounts

Counting Quarters

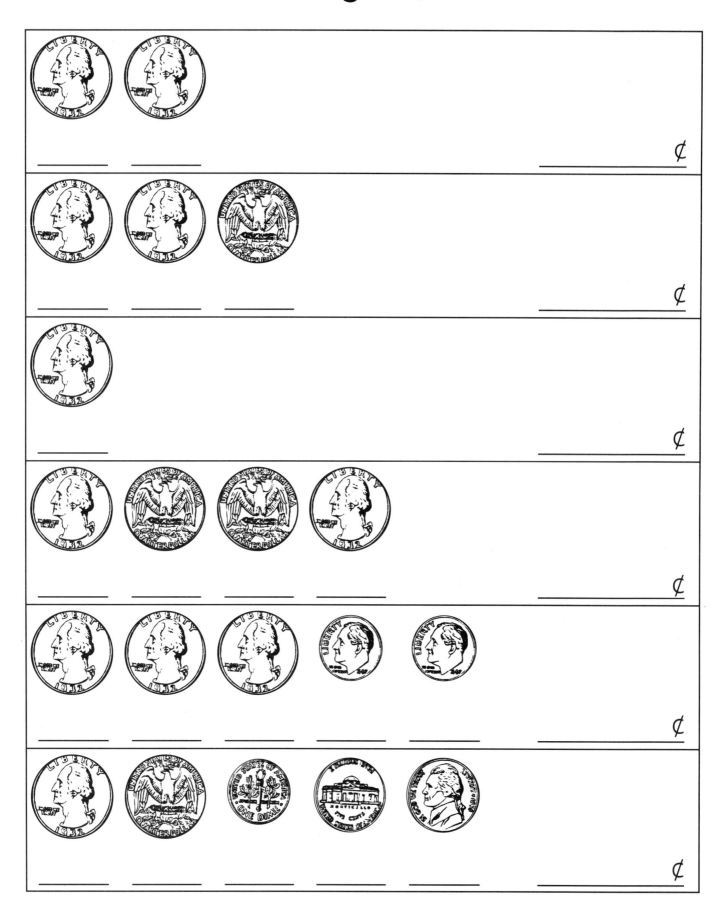

Greater than

Less than

Equal to

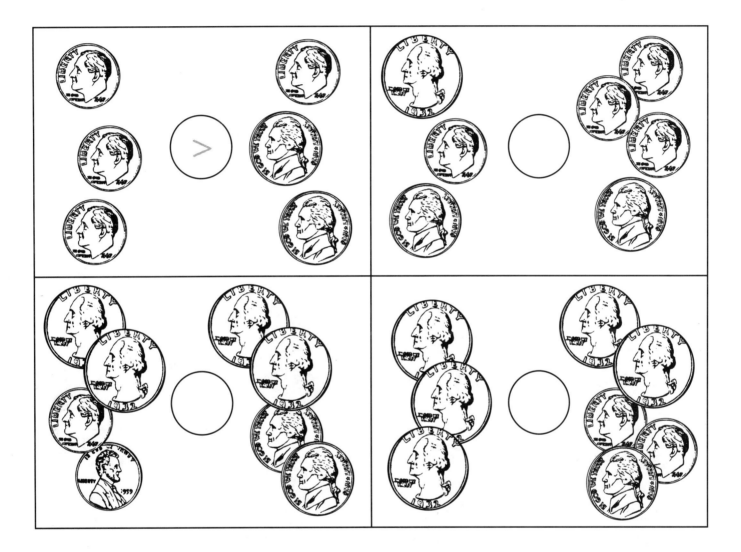

27¢ < 72¢   57¢ ◯ 97¢   92¢ ◯ 72¢

36¢ ◯ 41¢   83¢ ◯ 38¢   49¢ ◯ 49¢

49 Recognizing relationships between groups of coins

Color sets of 50 cents red.
Color sets of 25 cents blue.
Color sets of 5 cents yellow.

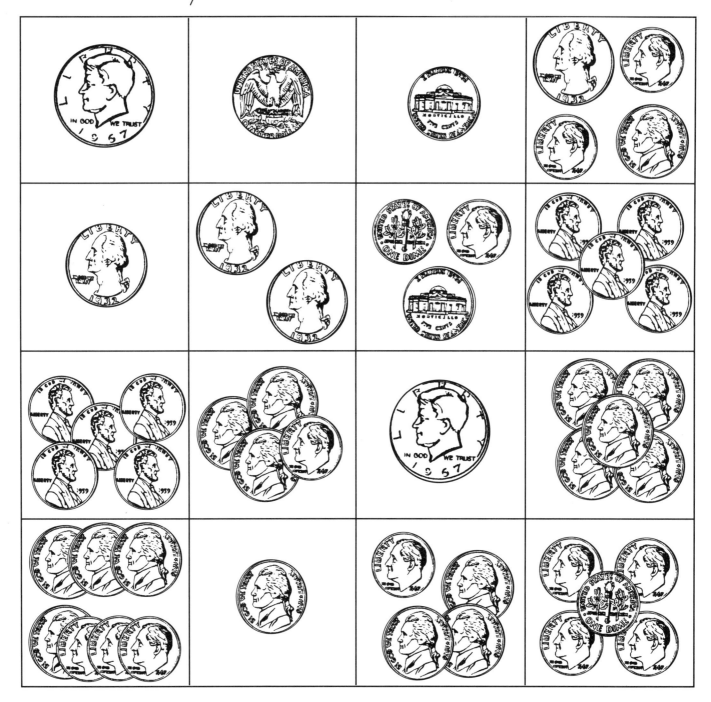

Match the coins to the amount.

a.

30¢

60¢

b.

27¢

50¢

c.

65¢

75¢

d.

35¢

43¢

e.

f.

g.

h.

51 Matching coins with amounts

Count to $1.00

1. $ .50 $ 1.00

2. $ __.__ $ __.__ $ __.__ $ __.__

3. $ __.__ $ __.__ $ __.__ $ __.__ $ __.__
   $ __.__ $ __.__ $ __.__ $ __.__ $ __.__

4. $ __.__ $ __.__ $ __.__ $ __.__ $ __.__
   $ __.__ $ __.__ $ __.__ $ __.__ $ __.__
   $ __.__ $ __.__ $ __.__ $ __.__ $ __.__
   $ __.__ $ __.__ $ __.__ $ __.__ $ __.__

5. How many in $1.00? _____

   How many in $1.00? _____

   How many in $1.00? _____

   How many in $1.00? _____

   How many in $1.00? _____

Circle coins to make sets of $1.00.

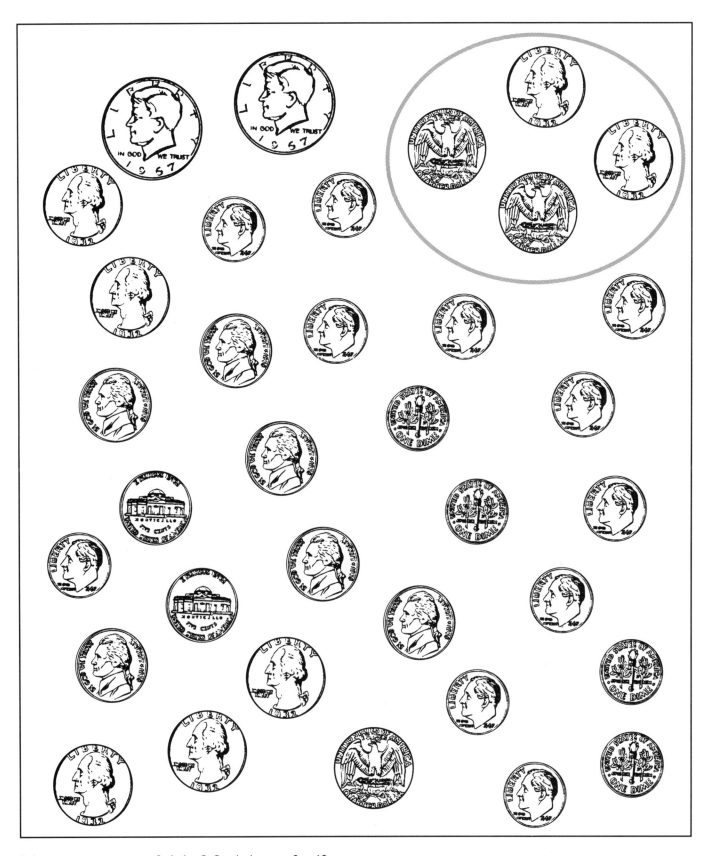

How many sets of $1.00 did you find?

Making Change

Costs:	You have:	Left:
🎈 32¢	quarter, dime, nickel, penny, penny, quarter, dime, nickel, penny, penny	_____ ¢
🪁 89¢	dime, dime, dime, penny, penny, penny, quarter, nickel, nickel, penny, penny, penny	_____ ¢
53¢ Yum Gum	quarter, quarter, dime, nickel, nickel, penny, penny, penny	_____ ¢
🍦 71¢	quarter, dime, nickel, penny, penny, quarter, dime, nickel, nickel, penny, penny	_____ ¢
🖍 99¢	quarter, quarter, nickel, penny, penny, quarter, quarter, nickel, penny, penny	_____ ¢

54 Making change; understanding money word problems

Dollar Sign $ Decimal Point .

Put the dollar sign and decimal point where they belong.

187 = $1.87   210 = _____

479 = _____   107 = _____

333 = _____   598 = _____

Count the money

$ 1.20

$ __ . __

$ __ . __

$ __ . __

$ __ . __

$ __ . __

55 Writing dollar amounts

How much money do you see?

 $ 5.75

 $.

 $.

 $.

 $.

 $.

Helping Grandpa

Maggie helped her grandpa clean out his garage. He told her she could have any of the cans, soda bottles, glass, or newpapers she found. Maggie was delighted. When she sold what she had cleaned out of the garage, she had made a lot of money.

1. Maggie found 18 aluminum cans and 24 soda bottles. She got 3 cents for each of the cans and 10 cents for each bottle. How much money did she get?

   $ ____ . ____

2. She sold the newspapers by the pound. She got 12 cents for each of the 20 pounds of newspapers she carried away. How much did she get for the newspapers?

   $ ____ . ____

3. Finally, she sold a box of old glass jars. She got 7 cents a pound for the glass. She was surprised to find that the box of glass weighed 48 pounds. How much money did she get for the glass?

   $ ____ . ____

4. What was the total amound of money Maggie got for the bottles, cans, glass, and newspapers?

   $ ____ . ____

57 Understanding money word problems

How much does your name cost?

vowels	"tail" letters	all other letters
$.50	$1.00	$.25
a e i o u	g j p q y	

Sam	Leroy	Reggy
$ .25 .50 .25 ——— $.		

My first name	My middle name	My last name

58 Understanding money word problems

Sums and Differences

 $ 2.50

 + 1.32

$.

 $

 +

$.

 $ 6.60

 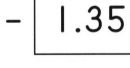 − 1.35

$.

 $ 16.53

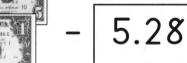  − 5.28

$.

$ 4.95 $ 8.31 $ 3.62 $ 8.64 $ 6.21
− 2.49   − 5.74   + 6.59   + 2.53   − 5.89
$ 2.46 $. $. $. $.

59 Adding money amounts

Making Change

You have:	Cost:	Give clerk:	Change:
(one dollar bill, two quarters)	$ .90	$ 1.00	$ .10
(one dollar bill, two dimes)	$ 1.30	$.	$.
(two one dollar bills)	$ 1.50	$.	$.
(two one dollar bills, quarter, dime, dime)	$ 2.28	$.	$.
(five dollar bill, one dollar bill, three dimes)	$ 5.08	$.	$.

60   Making change